Neil Armstrong
The First Man on the Moon

Dr Mike Goldsmith

HODDER
Wayland

an imprint of Hodder Children's Books

© 2001 White-Thomson Publishing Ltd

Produced by White-Thomson Publishing Ltd
2/3 St Andrew's Place, Lewes, BN7 1UP

Editor: Liz Gogerly
Cover Design: Jan Sterling
Inside Design: Joyce Chester
Picture Research: Shelley Noronha – Glass Onion Pictures
Proofreader: Alison Cooper

Cover: Neil Armstrong in 1969.

Published in Great Britain in 2001 by Hodder Wayland,
an imprint of Hodder Children's Books
This paperback edition published in 2002

Titles in this series:
Muhammad Ali: The Greatest
Neil Armstrong: The First Man on the Moon
Fidel Castro: Leader of Cuba's Revolution
Diana: The People's Princess
Anne Frank: Voice of Hope
Bill Gates: Computer Legend
Martin Luther King Jr.: Civil Rights Hero
Nelson Mandela: Father of Freedom
Mother Teresa: Saint of the Slums
Pope John Paul II: Pope for the People
Queen Elizabeth II: Monarch of Our Times
The Queen Mother: Grandmother of a Nation

British Library Cataloguing in Publication Data
Goldsmith, Mike, Dr.
    Neil Armstrong: the first man on the moon. -
(Famous Lives)
    1. Armstrong, Neil, 1930-
    2. Astronauts - United States
    I. Title II.
    629.4'54'092

ISBN 0 7502 3885 2

Printed in Hong Kong by Wing King Tong

Hodder Children's Books
An division of Hodder Headline Limited
338 Euston Road, London, NW1 3BH

Picture acknowledgements
The publisher would like to thank the following for giving
permission to use their pictures: Corbis (cover), 35, 43
(top and bottom), 45; Genesis (title), 7, 20, 22, 27, 28,
32, 36, 39, 40, 41, 42; Ohio Historical Society 8, 9, 10,
17; Photri 4, 5, 9, 13, 16, 18, 19, 21, 23 (top and
bottom), 24, 25, 26, 29, 30, 31, 33, 34, 37, 38, 44;
Popperfoto 6, 11, 12, 15; T.R.H /US Navy 14.

Contents

First Landing

The **Eagle** *leaving the Moon's surface. The Earth can be seen over the lunar horizon.*

It was 20 July 1969, and a spacecraft was approaching the surface of the Moon. It was called the *Eagle*, and it was one of the last remaining pieces of an enormous machine that had blasted off from the Earth three days before. For most of its journey, the ship had coasted through space under the influence of the gravity of the Moon and the Earth. The final approach to the Moon had been mostly computer-controlled, but now it was time for Neil Armstrong to take over.

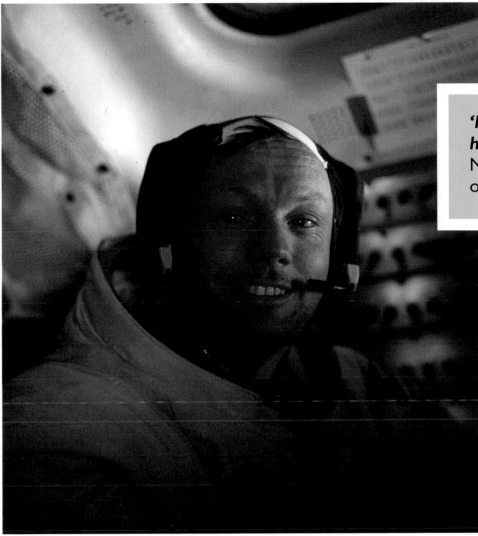

Neil Armstrong in the Eagle's cabin, having just returned from his exploration of the Moon's surface.

As the Moon's surface loomed closer, Armstrong saw that they were heading straight towards a field of huge boulders. If they tried to land there, the ship would be destroyed and Armstrong and his fellow astronaut Buzz Aldrin would die on the Moon.

Calmly, Armstrong steered the spaceship away from the boulders, but wherever he looked he saw more impossible landing-sites. Finally, with only about a minute's worth of fuel remaining, he saw a clear space ahead. Touchdown was so gentle neither astronaut felt it. Armstrong switched off the engine, and he and Aldrin looked out through the windows at the airless wastes of the Moon.

Destination Moon

People have always been fascinated by the Moon. The first calendars were based on its regular cycle of changes from new to full Moon, and there are all sorts of legends about the Moon and what the markings on its face are. When the telescope was invented in the early seventeenth century, the Moon's surface could be seen properly at last. It looked like the Earth — in some ways. Mountains and plains could be clearly seen, but was there water or air or even life?

'**The Moon certainly does not possess a smooth and polished surface, but one rough and uneven, and, just like the face of the Earth itself, is everywhere full of vast mountains, deep valleys, and sinuosities.'**
Galileo's report of his first telescopic observations of the Moon in 1610.

The Italian scientist Galileo Galilei with one of the many telescopes he built.

Although the first spacecraft from Earth only reached the Moon in the 1960s, there have been stories about travelling to the Moon for more than a thousand years. The fictional travellers got there using strange sources of power – including the wind, sunlight and even migrating geese! There were scientific reasons to go to the Moon (like finding out where it came from) but the main fascination was just to reach a different world that seemed so temptingly close.

The Moon, nearly full, photographed from Apollo II.

Birth of an Astronaut

'I must have been a staunch aviation fan before I was even conscious of it.'
Neil Armstrong.

Neil Armstrong was born on 5 August 1930 in a small town called Wapakoneta, in the state of Ohio in the USA. His father Stephen was an accountant who moved from one town to another, taking his family with him – so Armstrong had many homes in different parts of Ohio. One of these was in Cleveland, which had its own airport. When he was only two, his parents took him there to watch the planes. He soon had a fascination with flying that was to last for the rest of his life and eventually take him to the Moon.

In 1936, the family were living in the town of Warren. In those days, very few civilians, and almost no six-year-olds, had the opportunity to fly. But Armstrong got the chance. One Sunday he and his father sneaked off from church to pay a visit to Warren Airport to see a plane nicknamed 'the Tin Goose'. The pilot offered them a ride and Armstrong flew for the first time. He loved it, and soon he was spending all his free time building model planes.

Neil Armstrong, aged two, already fascinated by planes.

Right **Neil Armstrong, aged four, with his sister June.**

Below **A Ford Trimotor – the type of plane in which Neil learned to fly.**

War in Europe

Armstrong's fascination with planes lasted, and he was determined to fly one. To do that, he had to take flying lessons, and to do *that* he needed to earn some money. Armstrong found a job helping out at a shop where he had to work for over 22 hours to afford a single lesson! But he was dedicated and kept up the work and the flying lessons.

After two years, on his sixteenth birthday, he got his licence. At school Armstrong worked hard at lessons, but he didn't spend all his time studying: he started a jazz band, in which he played the horn, and he also joined the scouts.

'He always had a goal to work on.'
Armstrong's high school teacher.

A young Neil Armstrong in his band uniform.

In 1939, the Second World War had broken out in Europe, and the USA was drawn into the conflict in 1941. At the time it had had little effect on Armstrong, but he was soon to become personally involved in war.

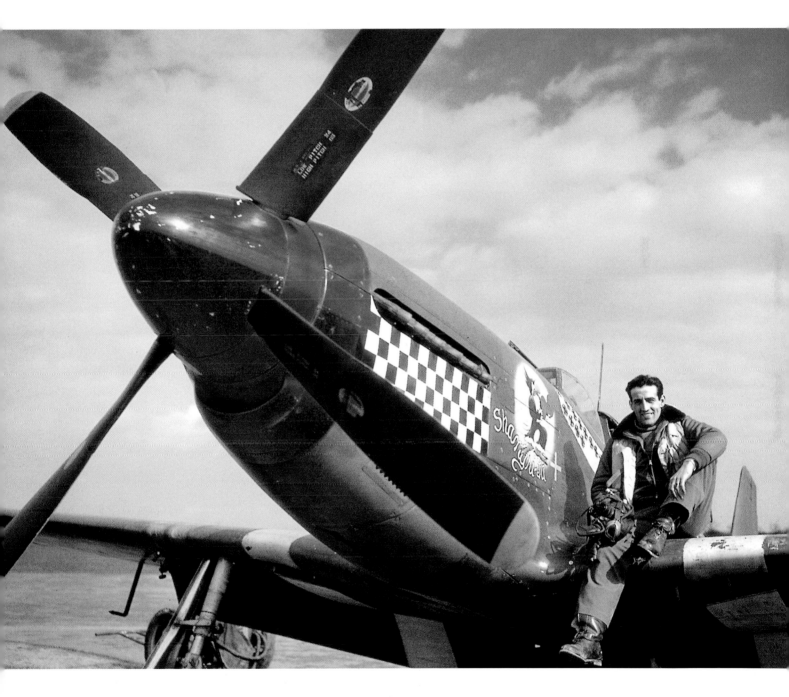

A Second World War American pilot poses on a Mustang in 1944.

Call-up for Korea

When he left school, Armstrong was determined to go to university to follow up his interest in aeronautics – the science of flight. The only problem was money – until the Navy offered him a scholarship. In 1947, he started a course in aeronautical engineering at Purdue University in Indiana, where he also joined the Naval Air Cadet programme.

'I was in the basement getting out quart jars of canned fruit to bake some pies. He called so loudly, 'Mom, Mom,' that he scared me to death. I dropped a jar of blackberries on my big toe. I must have broken the toe – it was black and blue for weeks. But that was such a great day.'
Viola, Armstrong's mother, remembering the day Armstrong got his scholarship.

A pilot of a B-29 plane prepares his crew for a bombing raid on Korea.

Soon after Armstrong started his university course, a war started in Korea. The USA fought with the South Koreans against the Chinese-supported North Koreans. The Navy selected young men with the skills they needed, and Armstrong was an obvious choice. In 1950 he was called up to fight. After some rapid training in Florida, he was assigned to an aircraft carrier – the USS *Essex* – as a fighter pilot. His dreams of being a flier had come dangerously true.

The USS Essex, the aircraft carrier from which Neil Armstrong flew over Korea.

Years of Danger

Armstrong flew 78 missions in the Korean War. Though he had trained in air-to-air combat, his job was to fly into enemy territory to damage or destroy bridges and other targets. It was dangerous work – to keep under cover, Armstrong had to fly very low, which required split-second reactions and excellent flying skills.

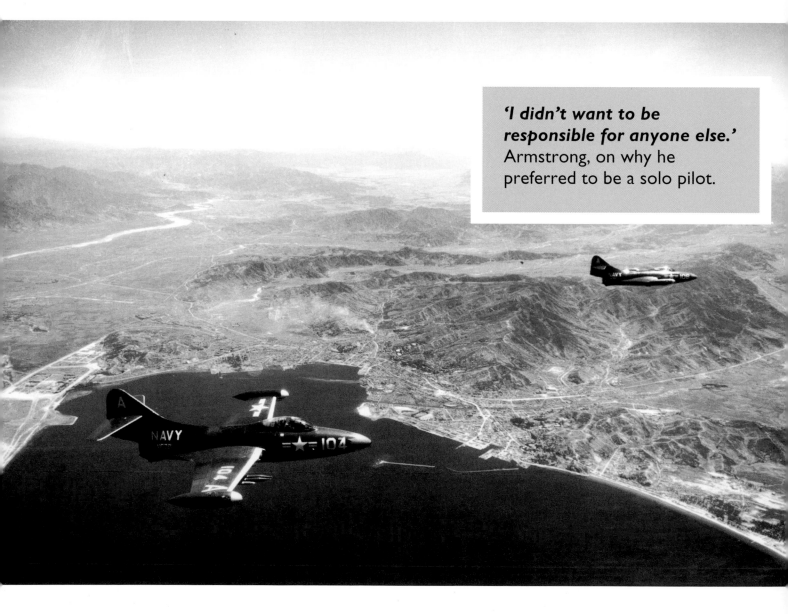

'I didn't want to be responsible for anyone else.' Armstrong, on why he preferred to be a solo pilot.

The type of plane Neil Armstrong flew in Korea – a Panther.

A US B-26 bomber drops a load of bombs on targets in North Korea in April 1950.

It wasn't just natural hazards that Armstrong had to face. Once, the tip of one of his plane's wings was sheared off by a cable placed in his path as a booby trap. It took all of Armstrong's flying skills to coax the damaged plane out of enemy territory, so that he could parachute to safety. On another occasion he managed to fly a badly damaged plane to a safe landing. Many of Armstrong's fellow pilots died in Korea.

In 1952, his tour of duty over, Armstrong returned to the USA to complete his university degree. He later received three medals for his courage in the war.

Falling in Love

Neil Armstrong at Edwards Air Force Base, during training in 1956.

'He flies an airplane like he's wearing it.'
One of Armstrong's fellow pilots at Edwards Air Force Base.

In 1955, Armstrong finished his university degree course and, still fascinated by flying, took a job as a pilot. But it was no ordinary flying job. He was posted to Edwards Air Force base in California, where a government organization called the National Advisory Committee for Aeronautics (NACA) was developing a new generation of rocket-propelled planes, which flew higher and faster than anything else in the sky.

In 1956, Armstrong married Jan Shearon. They had met four years earlier, at Purdue University, but it was a long time before they went out together and even longer before they married. As Jan said, 'Neil isn't one to rush anything.'

Armstrong has always been a very private person and he and Jan made a home in a forest cabin in the San Gabriel mountains in California, rather than in the town of Lancaster where most of his colleagues at the air force base lived.

Neil Armstrong proposed to Jan Shearon in the summer of 1955, when she was working at a summer camp in Winsconsin. They were married the following January.

Moon Race

In 1957, the Armstrongs had their first son, Eric. The Space Age was born the same year when the Soviet Union launched Sputnik, the first satellite, into space. The rest of the world was amazed. The US government and public were concerned that the Soviet Union had beaten them into space. They were even more worried when the Soviets put the first animal and the first human into space. The animal was a dog called Laika, and the man was the cosmonaut, Yuri Gagarin.

Yuri Gagarin, who at the age of twenty-seven became the first man in space on 12 April 1961.

The USA was determined to catch up, and in 1961 President John F Kennedy announced his determination to land an American on the Moon by 1970. It was a bold decision. At the time there had been only one American astronaut, Alan Shepard, and he had spent just fifteen minutes in space.

Alan Shepard, who became the first American in space on 5 May 1961.

'I believe that this nation should commit itself to achieving the goal, before this decade is out, of landing a man on the Moon and returning him safely to Earth. No single space project in this period will be more impressive to mankind, or more important for the long-term exploration of space – and none will be so difficult or expensive to accomplish.'
President John F Kennedy, May 1961.

The Edge of Space

In 1958, NACA changed its name to NASA, the National Aeronautics and Space Administration, and selected the first seven US astronauts to pilot a series of one-man Mercury capsules into space. Neil Armstrong was not among those who volunteered. For one thing, astronauts then were more like passengers than pilots. Pilots of rocket-planes, on the other hand, had to use all their skills on their high-speed, high-altitude missions. On one flight in 1962, in the X-15 rocket-plane, Armstrong reached 63,100 metres above the Earth — the edge of space. In total, Armstrong flew over a thousand hours in experimental rocket-planes like the X-15.

Neil Armstrong in the cockpit of the X-15 experimental rocket-plane.

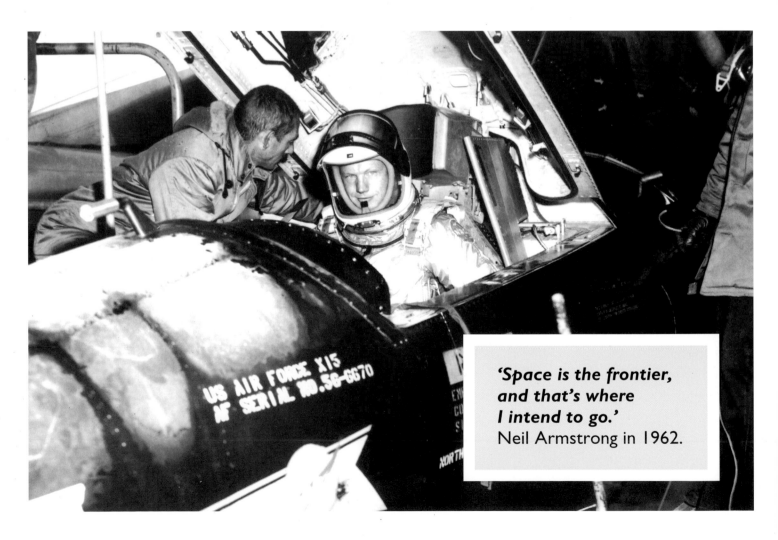

'*Space is the frontier, and that's where I intend to go.*'
Neil Armstrong in 1962.

Armstrong was to have piloted a new type of rocket-plane, bigger and faster than the X-15, called the Dyna-Soar. However, while this new rocket-plane was still on the drawing board, priorities changed. NASA's main goal was no longer to send rocket-planes to the limits of the atmosphere – it was to send a man to the Moon. When the next opportunity came, Armstrong volunteered as an astronaut – and was accepted.

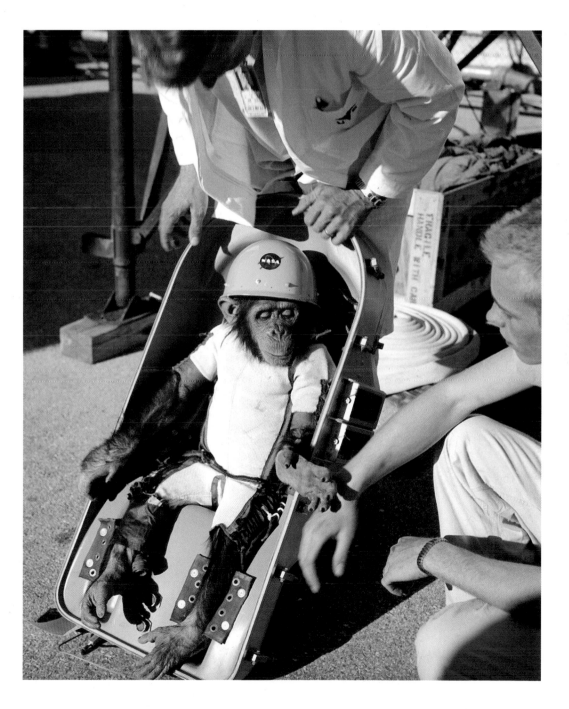

Before risking human astronauts, NASA sent animals into space, including a chimp called Ham. Like most test animals, Ham returned safely to Earth.

Training for Space

The crew of **Gemini 8** – Dave Scott and Neil Armstrong – pose for publicity photographs before their mission.

There was a great deal to learn about space travel before it would be possible to travel to the Moon, so NASA designed and launched a series of spacecraft to prepare for lunar missions. The first craft had no people on board, but later models carried two astronauts. These were called Gemini spaceships because Gemini is a constellation of stars which resembles a pair of twins.

Armstrong's first trip into space was as a Gemini astronaut, but first he had to go through some tough training. He was spun at high speeds to simulate the accelerations he'd have to cope with. He spent hours in a model of a Gemini cockpit, learning exactly how to fly it, and how to cope if things went wrong. He was even taught how to survive if his spacecraft came down in a jungle.

Armstrong was a member of the backup crew for *Gemini 5*, and then he was selected to be the commander of the *Gemini 8* mission in 1966. He had his chance to be a space-traveller at last. He didn't suspect how close he was to come to death in space.

> '**Their training had been fantastically intense; it had little to do with the public image of an astronaut jogging on the beach.**'
> From Gene Farmer's *First on the Moon*.

Above **Dave Scott and Neil Armstrong (background) learning to cope with the weightless conditions they would experience in space.**

Left **Neil Armstrong, taking his temperature as part of a medical check in preparation for his Gemini mission.**

Danger in Space

Armstrong's first space mission took place in 1966 with fellow astronaut Dave Scott. Their job was to join together two spacecraft in space, an operation known as docking, which had never been attempted before.

The launch of Gemini 8.

The experiment worked perfectly – Armstrong and Scott gently guided their spaceship to a special satellite, and then locked the craft together. But then, when they were 300 kilometres above the Earth, the linked spaceships began to tumble. Whatever Armstrong did had no effect, and the spin got faster and faster. He separated the two ships, but things only got worse, and soon *Gemini 8* was turning over every few seconds, and still getting faster. If Armstrong didn't do something soon, the ship would disintegrate and they would die in space.

'The Gemini spacecraft stopped responding to the controls and rotated more rapidly than ever – the sun flashed through the window about once a second. The sensations were much like those you would feel during an aircraft spin.'
Neil Armstrong.

*The Agena spaceship photographed from **Gemini 8** before docking.*

Armstrong had one more try   he switched off all the steering thrusters, and switched on the re-entry thrusters in the spaceship's nose. And it worked! Gradually the spinning slowed and stopped.

Back on Earth the cause of the problem was soon discovered – one of the steering thrusters had jammed on. If Armstrong hadn't switched them all off, his first space mission would have been his last.

A Tragic Beginning

None of the Apollo crews died in space, but tragically the astronauts who were to have been on the first Apollo mission died in 1967 when their capsule was being tested. They were killed by a fire that swept through the capsule in seconds. The astronauts were trapped because the hatch, which opened inwards, was held shut by the high air pressure in the capsule.

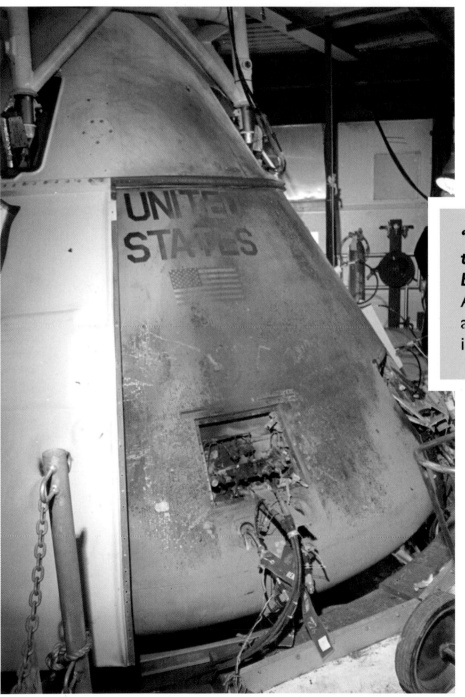

'If we die, we want people to accept it. We're in a risky business.'
Astronaut Gus Grissom, a few weeks before his death in *Apollo I*.

The scorched remains of the Apollo I *capsule*.

Armstrong himself had another narrow escape in 1968. This time he was flying the Lunar Landing Research Vehicle, an aircraft which was built to behave like the Lunar Module that would eventually land on the Moon. When Armstrong was flying the machine for the twenty-first time, and it was over 30 metres in the air, it started to tilt sharply to one side. As the aircraft went completely out of control Armstrong managed to eject himself just in time. While it crashed to the ground and exploded, he landed safely by parachute.

Neil Armstrong in a simulation of the cabin of an Apollo capsule.

The Moon-Landing Mission

The Apollo astronauts were assigned to their missions in strict rotation. Armstrong's first assignment was as a backup crew-member of the *Apollo 8* mission, in case any of the astronauts needed to be replaced. Then, in January 1969, he was assigned to *Apollo 11*, the Moon-landing mission. Soon after, Armstrong was told that he would be the first person to step onto the Moon. His fellow crew members were Buzz Aldrin and Mike Collins. Aldrin and Armstrong were to travel down to the Moon's surface in the Lunar Module while Collins was to remain in orbit around the Moon in the Command Module.

The crew of **Apollo 11**: *Neil Armstrong, Mike Collins and Buzz Aldrin.*

The next few months were a period of intense activity for Armstrong, Aldrin and Collins, as they spent up to fourteen hours a day practising on 'simulators' – imitations of the cockpits of the Lunar and Command Modules. There were also hours of lectures on rocketry, astronomy and geology, and many medical tests and interviews.

> **'What I really want to be, in all honesty, is the first man back from the Moon.'**
> Neil Armstrong.

Neil Armstrong and Buzz Aldrin training for their exploration of the Moon.

The Moonship

The spacecraft that would take Armstrong and his crew to the Moon was enormous, very powerful, and extremely complicated – and it also had to be very reliable.

The spaceship was composed of eight main parts, seven of which would be discarded at different stages of the journey. This made the spaceship lighter and therefore meant it needed less fuel.

The three stages of the *Saturn 5* rocket were mainly fuel tanks and rocket motors, used to drive the spaceship into space and start it on its journey to the Moon.

A diagram of the Saturn 5 rocket.

escape tower

service module

command module

lunar module (in two parts)

third stage

second stage

first stage

exhaust burners

APOLLO
LUNAR LANDING MISSION PROFILE

CSM-LM RENDEZVOUS
TRANSEARTH INJECTION
LIFTOFF
CSM TRANSEARTH TRAJECTORY (55-60 HRS.)
INSERTION
LM ASCENT
LUNAR ORBIT RETURN
CM WATER RECOVERY (PACIFIC)
LAUNCH
60 N. M. ORBIT
LUN. SURF. EXPERIMENTS
LUNAR ORBIT INSERTION
TOUCHDOWN
DESCENT
ASCENT
BEGIN LUNAR ORBIT
CSM TRANSLUNAR TRAJECTORY (65-75 HRS.)
TRANSLUNAR INJECTION

NASA HQ FP69-16541 6-11-69

When the escape tower and the three stages had been used and jettisoned, the three Apollo Modules would be left. The Lunar Module was the only half that would actually land on the Moon, and only the upper part of it would take off again. The job of the Service Module was to complete the journey to the Moon, orbit it and then take the astronauts most of the way home. It too would finally be discarded and only the Command Module would return to Earth.

A diagram of the Apollo II lunar-landing mission.

'It is a monster, that rocket.
It is not a dead animal; it
has a life of its own.'
Apollo rocket scientist
Gunter Wendt.

31

Escape from Earth

The most amazing week of Armstrong's life started early on July 16 1969. He got up at 4.15 am, and had a breakfast of steak and eggs. By 6 am he was being helped into his space-suit. A van took him, with Aldrin and Collins, past cheering photographers to the base of the rocket that towered 110 metres up into the morning sky. A lift took the crew to the top of the rocket where they swung through the narrow hatch. After checking the control systems, they had little more to do, except wait.

Neil Armstrong suited-up and getting ready for take-off.

The launch of Apollo II, with Neil Armstrong, Mike Collins and Buzz Aldrin on board.

At 9.32 am the Saturn's rockets fired with a deafening roar and a gush of bright yellow flames. Very slowly it started to lift from the launch pad, then gathered speed rapidly and raced up into the sky. Forty seconds later it was going faster than sound, and only 2.5 minutes after launch the huge first stage fell back to Earth. It was followed just six minutes later by the second stage. The rocket's third stage took it right out of the atmosphere and sent it into orbit 184 km above the Earth. The first part of the journey to the Moon had gone perfectly.

'At my window right now I can observe the entire continent of North America.' Neil Armstrong.

A Voyage through Space

The spaceship didn't orbit the Earth for long. Less than three hours after take-off the third-stage rocket fired again, launching the ship away from the Earth and towards the Moon at 38,800 km per hour.

Except when the rockets were fired, everything on the ship was weightless. Nothing fell, everything that wasn't fixed down drifted round the cabin, including the crew.

The Command and Service Module and Lunar Module abandon the third stage of the Saturn 5.

After a tricky manoeuvre to realign the Apollo Modules, the third stage was abandoned and the rest of the ship continued on towards the Moon, rotating slowly to avoid any parts being in the Sun too long and overheating.

Neil Armstrong in a simulation of the Apollo 11 cabin.

On board, the crew were in constant contact with Mission Control on Earth. Cameras on board the ship could even transmit pictures back to Earth. All over the world, hundreds of millions of people looked in on Armstrong and his crew.

Apart from a single correction to the flight-path, the rocket travelled unpowered towards the Moon. It gradually slowed down as the Earth's gravity tried to pull it back – until 62.5 hours after launch, when the Moon's gravity took over. Then the spaceship started to pick up speed again for the final stage of its journey.

Journey's End

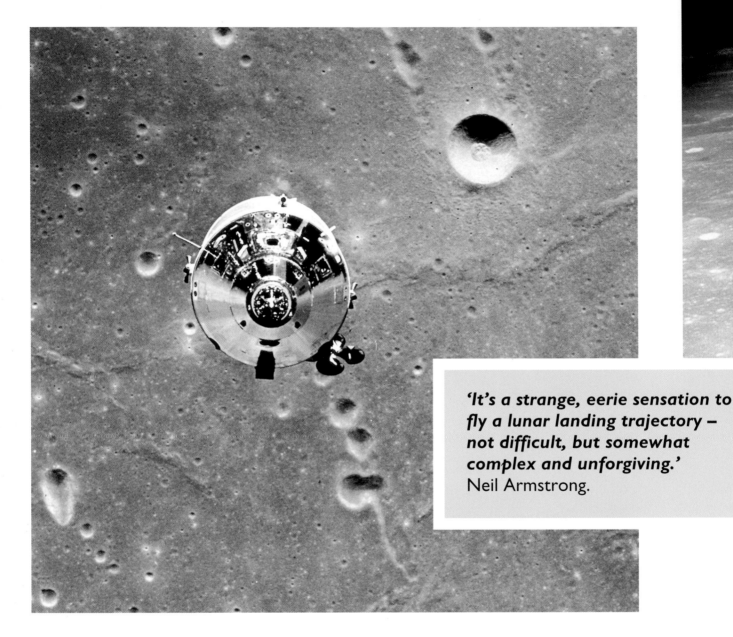

'It's a strange, eerie sensation to fly a lunar landing trajectory – not difficult, but somewhat complex and unforgiving.'
Neil Armstrong.

Before Neil Armstrong's lunar mission, the Apollo 10 astronauts carried out a test-flight and achieved lunar orbit in the Charlie Brown Command and Service Module, just as Columbia would.

Three days after launch, the spaceship's rocket motor fired briefly, changing its course to bring it into orbit around the Moon. After a second course adjustment, the ship was only a hundred kilometres above the lunar surface. Looking out, Armstrong and the others could see ranges of ancient mountains, great plains of dust and, wherever they looked, craters of all sizes.

The next stage was to prepare the Lunar Module – now called the *Eagle* – for its landing on the Moon. It was filled with oxygen, power was transferred to it and its landing legs were swung into position. Finally, Armstrong and Aldrin sealed the hatch and powerful springs separated it from the Command and Service Module (now called *Columbia*). The rocket motors of the *Eagle* fired and the little spaceship began its descent.

The flight wasn't as straightforward as Armstrong had hoped. The computer showed that they were off course slightly, and as they neared the surface, several alarm signals sounded – there was too much information for the computer to cope with. Undaunted, Armstrong guided the *Eagle* down to a perfect landing. They were on the Moon.

The Apollo 12 *Lunar Module* Intrepid approaches the Moon's surface, carrying a new crew of astronauts for the second manned landing.

A New World

Neil Armstrong took this picture of Buzz Aldrin about to step on to the Moon.

The plan was that the astronauts would rest before leaving the *Eagle*, but they were too excited to wait. They sealed their space suits, allowed the oxygen in the cabin to escape and opened the hatch. Then Armstrong crawled carefully out of the *Eagle*, down the ladder, and stepped onto the surface of the Moon. A few minutes later, Aldrin joined him.

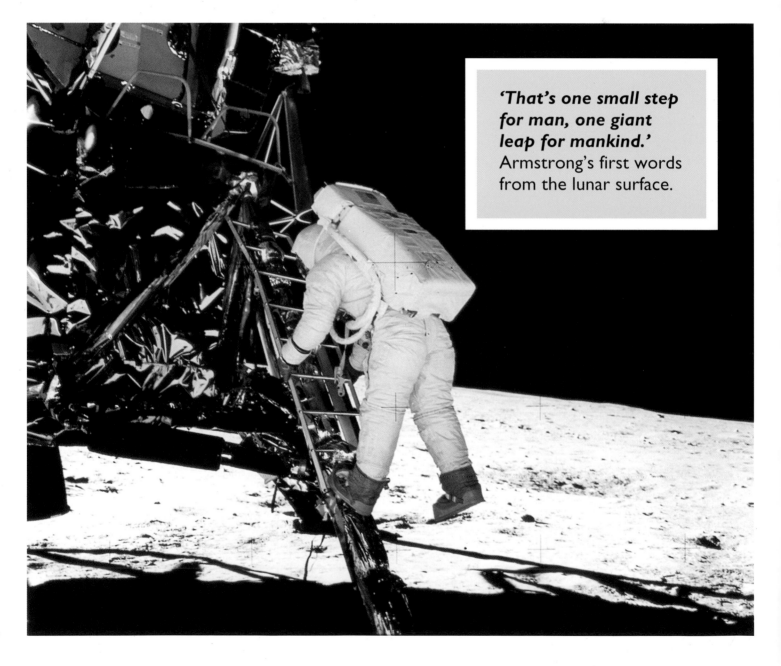

'That's one small step for man, one giant leap for mankind.' Armstrong's first words from the lunar surface.

The men had less than two hours to explore their new world — and they had plenty to do. They set up a lunar earthquake detector and a special type of mirror. A laser beam from Earth would be directed at it, to measure the distance to the Moon to within six centimetres. They experimented with different ways of moving around in the low-gravity environment of the Moon. They also collected rocks and dust, set up an American flag and spoke to Richard Nixon, the US President. Armstrong caught sight of some strange metallic globules and some clear crystals in the dust, but he had no chance to collect them. All this was watched on TV by more than one fifth of the world's population: over five hundred million people.

Buzz Aldrin exploring the Moon. The Eagle *is in* the background.

Return to Earth

A painting shows the upper half of the Eagle taking the astronauts back to Columbia. The lower half was left on the Moon.

All too soon, it was time to return to the *Eagle*. Armstrong and Aldrin climbed back up the ladder, filled the cabin with oxygen again and tried to sleep for a while. Then, 12 hours after landing on the Moon, it was time to leave. Armstrong triggered the rocket motor, and the *Eagle* separated into two halves — one half remained on the Moon while the other sped back up towards *Columbia* where Mike Collins was patiently waiting for them.

Three hours later, *Eagle* and *Columbia* docked – the manoeuvre that Armstrong had practised on his first, near-fatal space trip three years earlier. The astronauts were reunited, and the *Eagle* was allowed to fall back to the Moon, while *Columbia* started its long journey home. On the way, the astronauts were puzzled by some strange flashing lights in their eyes. Later, scientists decided the flashes were due to cosmic rays.

The Earth as seen from the Moon. For the astronauts it was a long way home.

'The responsibility for this flight lies first with history and with the giants of science who have preceded this effort.'
Neil Armstrong, on the journey back to Earth.

Splashdown

US divers met the Command Module after it splashed down into the Pacific Ocean. When leaving the Command Module, the astronauts wore special clothing to protect other people from any germs they might have picked up on the Moon.

After three days in space, *Columbia* was approaching the Earth again. After its rocket fired once more to correct their course, there was no further use for the Service Module, so it was jettisoned. Now only the flying-saucer-shaped Command Module was left to return the astronauts home. But if it flew directly into the Earth's atmosphere, it would burn up. So it flew a zigzagging path, swooping through the atmosphere briefly to lose speed before skimming out again to cool down. When it had done this several times, it was moving slowly enough to survive the last stage of the journey. It raced down through the atmosphere, glowing red-hot. The astronauts had to endure the effects of enormous changes in speed – after being weightless for so long, they had to cope briefly with weighing six times more than usual.

Neil Armstrong, Mike Collins and Buzz Aldrin (left to right) look out through the window of their quarantine capsule.

Finally, the parachutes opened and the Command Module splashed down into the Pacific Ocean. An hour later a helicopter was taking it back to the United States with Armstrong, Aldrin and Collins still on board. The first men on the Moon had returned to Earth.

'Neil, Buzz and Mike, I want you to know that I think I'm the luckiest man in the world, and I say this not only because I have the honour to be President of the United States, but particularly because I have the privilege of speaking for so many in welcoming you back to Earth.'
President Nixon, July 1969.

Jan Armstrong, with sons Eric and Mark, facing an army of reporters.

After Apollo

For 21 days after returning to Earth the astronauts had to stay in quarantine in case they had picked up any alien germs on their trip. Armstrong, Aldrin and Collins then went on a world tour, visiting 28 cities in 25 countries. Among the honours Armstrong received, one of the strangest was being sculpted in 400 kg of butter by the Ohio Dairymen's Association.

Armstrong has never been keen on publicity, and after the tour he did his best to avoid it. He rarely grants interviews and only occasionally makes public appearances. After working for a few more years at NASA, he became Professor of Aerospace Engineering at the University of Cincinnati, Ohio, and bought a farm nearby. He left the University in 1979.

The City of New York welcomes the returning astronauts with a ticker-tape parade.

Neil Armstrong at a ceremony in 1999 to collect the Langley Medal, which is awarded for the application of science to flight.

In 1986, the US Space Shuttle *Challenger* exploded soon after take-off, killing its crew of seven. Armstrong was vice-chairman of the enquiry to establish the cause of the disaster.

Armstrong is now chairman of AIL Systems Incorporated, a New York electronics company, but he still lives in Ohio. The footprints he left on the Moon will be there for centuries.

'I don't want to be a living memorial.'
Neil Armstrong.

Glossary

Aeronautics The science of air flight.

Air-to-air combat When an aircraft fights other aircraft in the air.

Assigned To be given a task or a position.

Atmosphere The layer of gases which surrounds the Earth.

Booby trap A harmless-looking device which can kill or injure anyone who touches it.

Boulders Large rocks.

Civilians People who aren't in the armed services or the police force.

Coasted Travelled without engine power.

Constellation A pattern of stars.

Cosmic rays Mysterious high-energy radiation found in space.

Cosmonaut A Soviet astronaut.

Crater A round hollow with a raised rim.

Discard To get rid of something when you have finished with it.

Disintegrate To crumble away into fragments.

Docking Joining two spacecraft together in space.

Eject To quickly leave an aircraft in an emergency.

Gravity The force that keeps people on the ground, makes stones fall to Earth and keeps the Moon orbiting the Earth and the planets orbiting the Sun.

Hatch A door or opening in an aircraft or a spacecraft.

Jettison To get rid of something which is no longer needed.

Lunar To do with the Moon.

Mission Control The group of people on the Earth who monitor space missions and remain in contact with the astronauts throughout.

Orbit The path of one object around another in space.

Quarantine A set period of time when a person is kept away from other human beings in case they have been infected with diseases that can be passed on.

Rotation When people in a team take it in turns to be on duty on a regular basis.

Satellite A natural or artificial object orbiting a planet.

Simulate To imitate conditions so closely that the simulation seems like the real thing.

Sinuosities Winding shapes.

Space shuttle A reusable spacecraft, which is launched with a rocket but lands like a plane.

Thruster A small rocket used to steer a spacecraft.

Touchdown The moment an aircraft or spacecraft lands.

Volunteer To offer your services of your own free will.

Date Chart

1930, August 5 Neil Alden Armstrong born in Wapakoneta, Ohio, USA.

1936 First flight.

1946 Obtains pilot's licence.

1947 Starts degree course at Purdue University.

1950 Called up to fight in Korea.

1952 Returns to Purdue University.

1955 Graduates with degree in Aeronautical Engineering.

1956 Marries Jan Shearon.

1957 Son Eric born. Sputnik launched. Armstrong becomes a test pilot.

1959 Daughter Karen born.

1961 US President Kennedy announces plan to land a person on the Moon by 1970.

1962 Becomes a trainee astronaut.

1963 Son Mark born.

1966 First space mission, as commander of *Gemini 8*.

1969 Becomes first human on Moon.

1971–1979 Works as Professor of Aerospace Engineering at University of Cincinnati.

1986 Vice-chairman of *Challenger* enquiry.

Further Information

Books to Read

A Man on the Moon by Andrew Chaikin (Penguin Books, 1995)

Gagarin and Armstrong: the First Steps in Space by Clint Twist (Evans Brothers, 1995)

Journey into Space by Andrew Langley (Kibworth Books, 1994)

Neil Armstrong by Carmen Bredeson (Enslow Publishers Inc, 1998)

Neil Armstrong by Sean Connolly (Heinemann, 1999)

Neil Armstrong: Young Flyer by Montrew Dunham (Aladdin Paperbacks, 1996)

One Giant Leap by Tim Furniss (Carlton Books, 1998)

Websites

NASA website www.nasa.gov

CD-ROM

Interfact Space Travel (Two-Can, 1999)

Index

Page numbers in **bold** mean there is a picture on the page.